My Artistic Conception Practised in Chang'an - Selection of Zhang Jinqiu's Architectural Creation

YAN'AN REVOLUTION MONUMENT

—此作品集已纳入中国建筑工程总公司《张锦秋建筑创作道路与思想成果集成研究》—
This volume has been included in "the Study on Collection of Achievements of Zhang Jinqiu's Architectural Creation Road and Thoughts" by China General Building Construction Company.

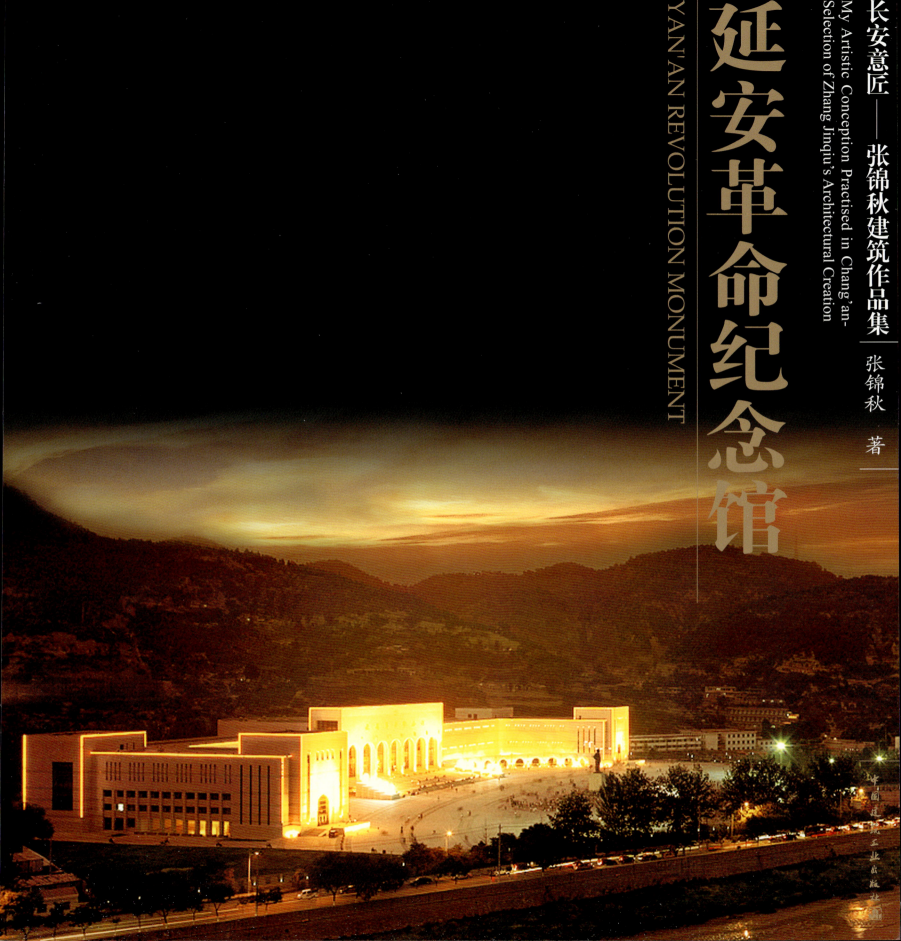

延安革命纪念馆

长安意匠——张锦秋建筑作品集

YAN'AN REVOLUTION MONUMENT

My Artistic Conception Practised in Chang'an—
Selection of Zhang Jinqiu's Architectural Creation

张锦秋 著

中国建筑工业出版社

目录 Contents

编者的话	5	Editor's Words
张锦秋简历	8	Resume of Zhang Jinqiu
代前言	11	Preface:
——和谐建筑之探索		Exploration on Harmonious Architectures
纪念性空间体系的营造	15	Creation of Memorial Space System
设计图	28	Design Drawings
实景照片	38	Scenes Photos
大环境	38	Environment
建筑形体空间	48	Building Shape Space
建筑主体形象	68	Main Building
建筑立面特写	78	Building Elevation Detail
室内	88	Interiors
后记	120	Postscript

长安意匠——张锦秋建筑作品集

编者的话 | Editor's Words

■自2006年3月~2008年7月，承编张锦秋院士《长安意匠——张锦秋建筑作品集》的各卷《圣殿记》《大唐芙蓉园》《现代民居群贤庄》《物华天宝之馆》陆续出版后，在城市建筑及文化各界引起强烈反响。作者张锦秋强烈的社会责任感和历史使命感，以及她旺盛的设计原创力，使她用好建筑为城市留下美好记忆，用作品集书写下能号召灵魂、启蒙创新观的著作系列。该系列作品集靠大气恢宏的建筑作品、严谨耐读的著述、优质的图文编撰品质已成为中国建筑出版界的"标杆"之作。鉴于该作品集后续各卷编撰已启动，本刊再撰新版"编者的话"。

这是一套展现东方建筑美学，开创独特创作风格的作品集；

这是一套凝聚世界建筑视野，扎根华夏睿智文化创新理念的思想库；

这是一套书写人生与城市、事业与国家、成就大师佳话的"传记"教科书。

■张锦秋20世纪60年代师从梁思成、莫宗江教授，1966年始投身西北建筑设计研究院，创作主持了数十项获国内大奖的项目。她是我国首批15位全国建筑设计大师中唯一的女性；1994年当选中国工程院首批院士；2001年获首届"梁思成建筑奖"；2005年7月获西安市委、市政府颁发的首届科技杰出贡献奖；2010年10月成为获"何梁何利基金科学与技术成就奖"的首位女科学家；2011年再获陕西省科学技术最高成就奖……面对如此多的殊荣，给我们留下最大教益的是她内心的谦和与淡定，是她与三秦父老、与西安这座历史文化名城难以割舍的情结，是她不懈为中国建筑文化传承与发展所表现的创新精神。

■张锦秋的建筑生涯是丰富的，她的建筑创作探索是多元的，她是中国罕见的能传承"唐风"建筑风格，能将中西方建筑语汇和谐运用，并能用最简洁的技法赋予建筑肌理以当代生命的文化使者。她关于历史文化名城视角下的建筑创作观及21世纪中国建筑设计发展方向的探索及成功经验，使她作品的光辉永驻，使她的贡献成为众多建筑后学的"榜样"。在《长安意匠——张锦秋建筑作品集》的全系列为读者已经展示的作品中，其代表性的创作理念解读扼要地归纳如下：

■体现"大象无形"境界的黄帝陵祭祀大殿。凡到过此地或阅读该书者，都仿佛进入"山水形胜、一脉相承、天圆地方、大象无形"的超凡境界。设计者为创造出雄伟、庄严、肃穆、古朴的炎黄子孙精神故乡的圣地感，在规划、格局、风格上都体现了传统与现代的气息，极其充分地展现了对文化遗产的尊重。轩辕殿的超大14m直径的圆形天光是最能诠释祭祀文化的现代元素，雨水、蓝天、白云、阳光都可无任何阻碍地进入祭祀大厅。在这里，人与建筑都融入山川形胜，不仅实现了人、建筑、自然的三位一体，更体现了天、地、人的高度融合。

■体现盛唐皇家园林文化的大唐芙蓉园。张锦秋表示大唐芙蓉园设计以传承弘扬华夏文化为宗旨，努力体现当代建筑师对盛唐历史文化的向往及发自内心的尊崇。面对唐长安被毁、曲江芙蓉园完全没有遗址的情况，创作者硬是靠对大唐芙蓉园唐诗、地方史志的深度挖掘，获得了创作的依据和灵感。这样，大唐芙蓉园的设计基调定位在体现唐代皇家园林宏大气势，并力求散发盛唐文化的文明感召力，使古为今用，服务当代，使每位步入佳境的宾客都有"走进历史、感受文明"之意，这是大唐芙蓉园成功创作的生命力所在。在此充分展示了其创作学养、人格技艺。

■体现陕西悠久历史和灿烂文化象征的"新唐风"陕西历史博物馆。该馆被列为国家第七个五年计划重点工程，规模是国家第二大博物馆。国家计委的任务书明确要求它成为陕西悠久历史和灿烂文化的象征。面对这特殊的挑战，张锦秋所在的中国西北设计院一共出了12个方案，有四合院、下沉式的现代建筑、窑洞等，而唯有她创作了一组唐代风格宫殿格局的现代建筑，获得了最终认可。对此张锦秋解读道：唐代是陕西历史发展的顶峰，而宫殿建筑集中体现了国家那个时代规划设计能工巧匠的最高水平。今天阅读陕西历史博物馆之所以仍感到它充满新意，还因为它有符合海内外不同参观者审美意趣的匠心设计。难怪在21世纪初西安评选的市民心中"新八景"中，共有张锦秋三项作品入选，"陕博"名列第二。2009年"陕博"被全国建筑业荣选入"新中国成立60周年百项经典工程"。

■伴随着2010年的结束，我们已完整走过21世纪的第一个10年。面对悠久的中国建筑艺术的长河，面对波澜不惊的世界级创作生态文化环境，我们又欣喜地发现，张锦秋的设计新品迭出：富于革命传统文化的延安纪念馆、满载盛唐记忆的大明宫丹凤门、即将迎来第41届世界园艺博览会的标志建筑"天人长安塔"……我们认为这些或感动、或震撼、或绝妙的精品之作，不仅会让四海宾朋纷至沓来，更会给中国建筑界留下串串思索：何为中国优秀建筑作品的成功创作途径？何为一代青年建筑师提升设计的城市文化品质的经验借鉴？何为实践一位设计大师与一个城市建设的完美结合？何为现代化理念下用作品与思想对东西方文化的论衡？愿海内外城市建筑规划设计者都能从她语淡、言真、意深的作品集中阅读感受到这一切，更希望通过不断省思，再激起对未来的更多期许。这些就是《建筑创作》杂志社走近张锦秋院士、承编她的著作集、向海内外同仁传播她的作品及思想之缘由。

《建筑创作》杂志社

2011年4月

■From March 2006 to July 2008 we edited "Story of Holy Temple", "Tang Lotus Garden", "Modern Folk Qunxian Manor" and "Museum of Treasures" by academician Zhang Jinqiu in the series of her "Artistic Conception Practiced in Chang'an – Selection of Zhang Jinqiu's Architectural Creation". Successive publications of these works gave rise to strong responses from the city architectural and cultural circles. Zhang Jinqiu's strong social responsibility, sense of historical mission and energetic design originality leave beautiful time for the city with buildings, inspires people's soul and enlightens their innovation consciousness with her works. Her special urban cultural qualification and persuasion make it possible for her to incorporate her designs with works. In her works we experience condensed words, harmonious atmosphere and noble ideals. She sets up a model for uncommon thinkers to advocate, culture masters to march and actors to harvest. The series of works has become a bench mark among the Chinese architectural publishers in the grandeur architectural designs, readable precise works and excellent drawings and specifications. Since the editing work of the successive volumes of the series has begun we write the new "Editor's words".

The series of works show that oriental architectural aesthetics presents a unique creation style.

The series of works is a crystallization of new ideas that combine the world architectural views with brilliant Chinese culture.

The series of works is a biography that explains her life and the city, her cause and the state, the career as a great master.

■ Zhang Jinqiu studied architecture under Professors Liang Sicheng and Mo Zongjiang. Since 1966 she has devoted in design researches on north-west China architecture and designed or taken charge of designing of a dozen projects which have been awarded grand prizes. She was the only one female architect of the first 15 national architectural design masters. In 1994 she was elected one of the first batch academicians of China Engineering Academy. She won the first Liang Sicheng Architectural Prize in 2001 and obtained the science and technology outstanding contribution prize first issued by the Xi'an municipality committee and government in July 2005. She was the first woman scientist who was awarded the science and technology achievement prize of Heliang Heli Foundation in October 2010 and the science and technology highest achievement prize of Shaanxi Province. With so many special prizes she shows modesty and indifferent. She has taught us very much about her deep affection for the history and culture of the well-known city Xi'an and about her creation spirit expressed in her untiring perseverance in carrying forward and developing Chinese architectural culture.

■ Zhang's architectural carrier is fruitful and her exploration on architectural creation is multi-principle. She is a rare cultural envoy who can carry on the Tang architectural style, use Chinese and western architectural languages in harmony and display architectural texture and modern life by means of simple techniques. Her view on architectural creation in historical and cultural cities, her exploration and successful experiences in the development of the Chinese architectural design in the 21 century make her works full of glory and present her contributions as architectural models. In the works already published for the readers in her Artistic Conception Practiced in Chang'an – Selection of Zhang Jinqiu's Architectural Creation we will read her typical creation concepts which we may outline as follows:

■ The grand sacrificial hall of the Yellow Emperor's mausoleum showing "grand sight having no definite form"

All those who have been to the sacrificial hall and those who have read the book will feel as if they entered an extraordinary place with beautiful maintains and rivers, continuous line and same origin, round sky and square earth, grand sight and indefinite form. The designer has planned a holy place that is full of elegance, dignity, solemnity and simplicity for the descendants of Yandi and Huangdi. Combination of tradition with modern atmosphere is displayed in planning, layout and style, which expresses the designer's respect to cultural heritage. The Xuanyuan hall has a super-large circular skylight of a 14 m diameter which is taken as a modern element to explain sacrificial culture. Rainwater, blue sky, white clouds and sunlight can reach into the sacrificial hall without any obstacles. Here, man and architecture are immersed in maintains and rivers in realization of man, architecture and nature combined in a whole and heaven, earth and man merged in harmony.

■ The Tang lotus garden showing the Tang imperial garden culture

In designing the Tang lotus garden Zhang Jinqiu followed the principle of Chinese culture and stated clearly the yearning and respect by the modern architects for the historical culture in the Tang dynasty. As the Chang'an city of the Tang dynasty had been destroyed and no relic site of the Qujiang lotus garden existed the designers gained basis and inspiration through assiduous study of the Tang poems on the Tang lotus garden and the annals of local history. The basic design goal of the Tang lotus garden is to express the grandeur of the Tang imperial garden, convey the inspiration of the Tang culture and make the past serve the present. Whoever comes into the garden has a feeling of "entering the history and experiencing the civilization". This is the vitality of the Tang lotus garden design and here we see her great attainments and perfect skills.

■ Shaanxi historical museum showing Shaanxi long history and resplendent culture

Shaanxi historical museum was the second biggest one of the same kind in the country's 7th five-year plan. In the design programme issued by the state planning committee it was clearly stated that the museum should become a symbol of Shaanxi long history and glorious culture. To face the challenge China North-west Design Institute worked out 12 schemes, of which some incorporated courtyards, some had sunk-type modern buildings or caves. Finally Zhang's scheme of Tang style imperial palace was accepted. She explained that the Tang dynasty was the peak period in Shaanxi history and imperial palace architecture expressed the highest level of planning and workmanship in that time. Today Shaanxi historical museum is still full of originality as it meets the aesthetic interests of visitors from home and abroad. It was natural that in the beginning of the 21st century Zhang's three design works were among "the Eight New Sights" appraised by Xi'an citizens, of which the museum was the second. In 2009 it was honoured with one of "the 100 classic projects in the 60 anniversary of the country".

■ The year of 2010 has passed and we have experienced the first 10 years of the 21st century. In face of the Chinese architectural art of a long history and the ecological and cultural environment of the design world on a magnificent scale we are pleased to find that new design works by Architect Zhang have been presented before us. The Yan'an memorial full of revolutionary tradition and culture, the Danfeng gate of the Daming palace in memory of the Tang prosperity and the "Tianren Chang'an Tower" which will be a landmark of the forthcoming 41st World Gardening Exhibition, etc. will make us moved or excited for their excellence. These projects will draw visitors from all over the world and leave the Chinese architectural circle for thoughts. Where can we find the successful way for Chinese excellent architectural design? What are the experiences for the young architects to upgrade their design level and improve the city cultural quality? How will a master incorporate his or her design with urban construction? How will one contain the east and west cultures in one's works and ideology? We hope that Chinese and foreign architects and city planners will find the answers from reading her series of works with plain words, true stories and thoughtful meanings and furthermore we want to arouse more thinking and inspiration for the future. This is the reason for us "The Architectural Creation" to approach Academician Zhang and edit the series of works and convey her design concepts and works to the colleagues of China and the world.

Architectural Creation Magazine Publishing House

April 2011

张锦秋简历 | Resume of Zhang Jinqiu

张锦秋

女，1936年10月生于四川成都。1954~1960年清华大学建筑系毕业，1961~1966年清华大学建筑历史与理论研究生毕业。1966年至今在中国建筑西北设计研究院从事建筑设计。1987年任院总建筑师，1988年晋升为教授级高级建筑师，1997年获准为国家特批一级注册建筑师，2005年当选亚太经合组织（APEC）建筑师。2010年任中国中建设计集团有限公司总建筑师。

主要获奖作品有：

阿倍仲麻吕纪念碑	1981年获国家建工总局优秀工程奖
陕西省体育馆	1986年陕西省优秀设计一等奖
法门寺工程	1991年建设部优秀设计表扬奖
	2009年新中国成立60周年中国建筑学会创作大奖
三唐工程	1992年获国家优秀勘察设计铜奖
	2009年新中国成立60周年中国建筑学会创作大奖
陕西历史博物馆	1993年获国家优秀勘察设计铜奖
	1993年获中国建筑学会首届建筑创作奖
	2009年新中国成立60周年中国建筑学会创作大奖
	2009年新中国成立60周年百项经典工程
西安钟鼓楼广场及地下工程	2000年获建设部优秀规划设计二等奖
大慈恩寺玄奘三藏法师纪念院	2002年获国家优秀勘察设计铜奖
西安国际会议中心、曲江宾馆	2003年获陕西省优秀设计一等奖
群贤庄小区	2004年获全国优秀勘察设计金奖
	2009年新中国成立60周年中国建筑学会创作大奖
陕西省图书馆	2004年获全国优秀勘察设计铜奖
大唐芙蓉园	2006年建设部优秀城市规划设计一等奖
	2009全国优秀工程勘察设计银奖
	2009年新中国成立60周年中国建筑学会创作大奖
黄帝陵祭祀大殿（院）	2009年全国优秀工程勘察设计金奖
	2009年新中国成立60周年中国建筑学会创作大奖
延安革命纪念馆	2009年新中国成立60周年百项经典工程

Zhang Jinqiu, female, born in October 1936 in Chengdu, Sichuan Province, studied in the Architecture Department of Tsinghua University in 1954-1960, majored in architectural history and theory for Tsinghua University postgraduate in 1961-1966. Since 1966 she has worked in China Northwest Architectural Design and Research Institute for architectural design. In 1987 she was appointed to chief architect of the Institute. In 1988 she was promoted to be a professor-grade architect and in 1997 approved to be a first class registered architect of the state. She was elected an architect of APEC in 2005. Chief architect of the chinese zhongjian design group co. In 2010.

Significant Awarded Design Works:

Monument to Abenonakamaro	1981, Excellent project prize of the State Construction General Bureau
Stadium of Shaanxi Province	1986, First Prize of Excellent Design of Shaanxi Province
Project of Famen Temple	1991, Praising Prize of Excellent Design of the Ministry of Construction
	Won the Grand Prize of Creation of the Chinese Architectural Society in Celebration of the 60th Anniversary of the People's Republic of China in 2009
Santang Project	1992, Bronze Prize of National Excellent Investigation & Design
	Won the Grand Prize of Creation of the Chinese Architectural Society in Celebration of the 60th Anniversary of the People's Republic of China in 2009
Shaanxi History Museum	1993, Bronze Prize of National Excellent Investigation and Design
	1993, First Architectural Creation Prize of Architectural Society of China
	Won the Grand Prize of Creation of the Chinese Architectural Society in Celebration of the 60th Anniversary of the People's Republic of China in 2009
	2009, One of the hundred classic in celebration of the 60 anniversary of the country
Square of the Bell and Drum Tower of Xi'an and its Underground Work	2000, Second Prize of Excellent Planning Design of the Ministry of Construction
Master Monk Xuanzang's Memorial Hall of Daci'en Temple	2002, Bronze Prize of National Excellent Investigation & Design
Xi'an International Conference Center and Qujiang Hotel	2003, First Prize of Excellent Design of Shaanxi Province
Modern Folk Qunxian Manor	2004, Gold Prize of National Excellent Investigation & Design
	Won the Grand Prize of Creation of the Chinese Architectural Society in Celebration of the 60th Anniversary of the People's Republic of China in 2009
Library of Shaanxi Province	2004, Bronze Prize of National Excellent Investigation & Design
Tang Lotus Garden	2006, First Prize of Excellent Planning Design of the Ministry of Construction
	2009, Silver Prize of National Excellent Investigation and Design
	Won the Grand Prize of Creation of the Chinese Architectural Society in Celebration of the 60th Anniversary of the People's Republic of China in 2009
Sacred Palace (Courtyard) of the Mausoleum of the Yellow Emperor	2009, Gold prize of national excellent investigation and design
	Won the Grand Prize of Creation of the Chinese Architectural Society in Celebration of the 60th Anniversary of the People's Republic of China in 2009
Yan'an Revolution Monument	2009, One of the hundred classic in celebration of the 60 anniversary of the country

鉴于张锦秋的学术贡献，1991年获首批"中国工程建设设计大师"称号、1994年被遴选为中国工程院首批院士，2001年获首届"梁思成建筑奖"，2004年获西安市首届科学技术杰出贡献奖，2010年获何梁何利科学与技术成就奖，2011年获2010年度陕西省科学技术最高成就奖。

In appreciation of Zhang's academic contribution she was entitled "Design Master of China Construction and Design" in the first batch in 1991, elected to be a member of the Chinese Academy of Engineering in the first batch in 1994, won the first Liang Sicheng Architectural Prize in 2001 and Outstanding Construction Prize of Science and Technology of Xi'an in 2004 and Heliang Heli Science and Technology Achievement Prize in 2010. Won the top achievement prize of science and technology of 2010 by shaanxi province in 2011.

长 安 意 匠 —— 张 锦 秋 建 筑 作 品 集

代前言——和谐建筑之探索
Preface: Exploration on Harmonious Architectures

世纪之交,东方正面临现代与传统、外来文化与本土文化的冲撞与融合。具有鲜明文化属性的建筑也不例外地卷入了这一浪潮。

At the turn of the centuries, the east world is facing the collision and fusion between the modern and the tradition, and between alien culture and local culture. Architecture, with its distinct cultural nature, is inevitably involved.

从哲学思潮来看,当代城市建设体现了科学主义思潮和人文主义思潮的汇合。在这个汇合点上,物质的与精神的、传统的与创新的、地域的与世界的等两极的东西必然会神奇般地统一起来,从而构成一种洋溢着生命气息和生活朝气的综合美。越来越多的建筑师认识到当代城市艺术的最大特征是综合美。这种美具有多元性和多层次性,其最重要的特性是和谐。

In the light of philosophic trends, contemporary urban construction reflects the converging of both scientism and humanism thinking, in which miraculous harmony is expected to be established between such opposite poles as the physical and the spiritual, the conventional and the innovative, the local and the global, and hence a synthetic beauty full of life and vitality is created. More and more architects realize that the most distinctive feature of modern urban arts is synthetic beauty. This beauty is of diversities and multiple strata with harmony as its most significant character.

建筑是人与人、人与城市、人与自然的中介,作为城市的主要组成,其文化取向当然应该与它所处的城市、环境相协调。优秀的建筑应该促进人与人的和谐,人与城市的和谐,人与自然的和谐。因此,我的建筑创作可以说是在追求一种"和谐建筑"。

Architecture is the medium between man and man, man and his city, and man and nature. As the main component of a city, the cultural orientation of architecture in a city should go with the city and the environment. Fine architecture is expected to promote the harmony between man and man, man and his city, and man and nature. In this sense, the endeavor in my career has been a pursuit of "harmonious architectures".

我在设计实践中,逐渐体会到"和谐建筑"的理念包含两个层次。第一个层次是"和而不同",第二个层次是"唱和相应"。中国古代哲人孔子说:"君子和而不同,小人同而不和"。"和"是指相异因素的统一,"同"是指相同因素的统一。我们赞赏前者,提倡"海纳百川,有容乃大",主张吸纳百家优长,兼集八方精义;第二个层次"唱和相应"是讲相异的因素怎么才能达到"和"的境界。古籍《新书·六术》上说:"唱和相应而调和"。这是讲音虽有高低不同,只要有主次,有节奏、有旋律地组织起来就可成为和谐的乐曲。先人的智慧给我们以启迪,有助于我们建筑师开扩设计思路,提高创作境界。在国际化的浪潮中,一方面勇于吸取来自国际的先进科技手段、现代化的功能需求、全新的审美意识,一方面善于继承发扬本民族优秀的建筑传统,突显本土文化特色,努力通过现代与传统相结合、外来文化与地域文化相结合的途径,创造出具有中国文化、地域特色和时代风貌的和谐建筑。

During my practice as a designer I have gradually come to realize that the idea of "harmonious architectures" consists of two strata. The first stratum is "diversified unification other than unified identicalness", the second stratum is "precenting and chorusing in unison". The ancient Chinese philosopher Confucius once said: "the noble seek diversified unification other than unified identicalness, while the mean follow the opposite to the

长 安 意 匠 —— 张 锦 秋 建 筑 作 品 集

uniformity of diversified factors while "identicalness" refers to the uniformity of identical factors. We agree with the former and promote the idea that "it is taking in all waters with all-embracing generosity that makes the immenseness of the ocean." We insist that we should adopt the essences and excellences of various schools and opinions. The second stratum "precenting and chorusing in unison" addresses how to achieve the harmonious 'unification' with these diversified factors. The Ancient Chinese book New Book • Six Skills reads: "precenting and chorusing in unison brings consonance of tunes." It explains that though there are different musical scales sounds they may be rhythmically and melodiously organized into consonant music with clear differentiation of the primary and secondary tunes. Wisdom of the forefathers gives us inspiration and contributes to the widening of our design vision and upgrading of creativity levels for architects. In the process of globalization, harmonious buildings with distinctive Chinese cultural, local and epochal features can be created by courageously adopting advanced technological means, modern function requirement and new aesthetic notion from abroad on one hand, and on the other hand, by carrying forward the excellent national architectural tradition and giving prominence to the local cultural features via integration of modernity and tradition, alien culture and local culture.

作为一名从业建筑师,我长期生活工作在古都西安。那是一座具有3100年城龄的古都,曾有13个王朝在这里建都。中华民族盛世王朝周、秦、汉、唐建都于此,达千年之久。这里是著名的丝绸之路的起点。西安是中华民族的精神故乡,至今保存着伟大的遗址、完整的城垣、重要的古迹,它们生动地述说着古都光辉的历史;西安现在是西部重镇,是我国现代化建设中西部大开发的中心城市,现代化建设正在迅猛发展。科技开发区、经济开发区、旅游开发区体现着当今城市的活力。就在这片古今交融、新旧相辉的热土之上,正在回荡起民族文化复兴的壮丽乐章。这一切成为我们进行建筑创作的广阔背景。"和谐建筑"的理念就由此而萌生。我深信和谐建筑所创造的物质环境和文化精神能够有利于增强民族文化认同感与归属感,有利于巩固和发展自身的社会凝聚力,在历史的长河中生生不息。

As an architect, I have been living and working in Xi'an for quite a long period. Xi'an is an old city with the history of 3100 years, which served as the capital of 13 dynasties. Over the past thousand years the prosperous dynasties of Chinese nation like Zhou, Qin, Han and Tang, all established their capitals here. It is also here that the famous Silk Road starts. Xi'an is the spiritual hometown of Chinese nation where still preserves the great historic sites, integral city walls, and important antiquities which vividly retell the glorious history of this ancient capital. Xi'an now is the center of the development of west China with its rapid growth in modern construction. Scientific development zones, economic development zones and the tourist development zones reflect the vitality of the city nowadays. It is on this hot land where the sublime music of national cultural renaissance resounds, with the antiquity and modernity blending, also with the old and new adding brilliance to each other. All these have composed a vast background against which we proceed with our architectural creation. The idea of "harmonious architectures" hereupon comes into being. I deeply believe that material environment and cultural spirit created by harmonious architecture will help enhance the feeling of recognition and destination of the national culture and consolidate and develop self social cohesion and they will last in the long river of history.

Zhang jinqiu

长安意匠——张锦秋建筑作品集

Creation of Memorial Space System

纪念性空间体系的营造

坐落在陕北茫茫黄土高原上的延安是中国革命历史上的一座重要里程碑。从1935年10月到1948年3月延安曾是中国工农红军两万五千里长征的落脚点，是十三年间中国共产党中央所在地，是抗日战争的政治指导中心，是中国共产党人集体智慧的结晶——毛泽东思想的诞生地，是延安精神的发源地，是新民主主义红色政权雏形的孵化地，是夺取全国胜利的出发点。为了展现中国共产党波澜壮阔而又艰苦卓绝的革命历程，让年青一代更好地接受爱国主义教育和革命传统教育，2004年6月国家决定在老馆原址上重建延安革命纪念馆。

Yan'an, located in the vast loess plateau of the north Shaanxi Province, was an important mile stone in the revolutionary history of China. In the years from October 1935 to March 1948 Yan'an was the stop place of the 25000 li Long March of the Chinese Workers and Peasants Army, the political instruction centre in the War of Resistance against Japan, the birthplace of Mao Zedong Thought which was the crystal of the collective wisdom of the Chinese Communist Party, the cradle of the Yan'an spirit, the hatching place of the initial form of a red regime of new democracy and the starting place for fighting for the complete victory in the country. To exhibit the unfolding and steadfast revolutionary history of the Communist Party and to educate the young generation with patriotism and revolutionary tradition, the state decided to renovate the Yan'an Revolution Monument on its original site in June 2004.

　　原延安革命纪念馆建于1969年，一直是延安革命历史的生动教材和旅游热点。由于遭受1977年山洪的浸泡，馆房已成危房，加之当时建筑的规模、标准、设施水平都比较低下，不敷使用，故而决定原址重建。延安革命纪念馆是延安当代标志性的纪念性建筑。它应该具有标志性建筑独一无二卓尔不群的品格，它应该浓缩延安精神的精华，传递着光荣的革命传统，标志着城市的灵魂和象征。这就给纪念馆的建筑设计提出了很高的要求。

The original Yan'an Revolution Monument was built in 1969. It became a vivid textbook and a popular tourist spot. As a result of soaking of the torrents in 1977 it was damaged seriously. In addition the construction regulation, standards and installation in that time were in low level. Renovation of the old Memorial Hall was a right decision. The Yan'an Revolutionary Memorial Hall is a modern landmark in Yan'an. It presents unique characteristics of a memorial building, expresses the crystal of the Yan'an spirit and transmits glorious revolutionary tradition. It is the soul and symbol of the city. All this raises high requirements for the architectural design of the memorial.

2004年12月我们从可行性研究入手承担了该项目的设计工作。之前在延安市领导的率领下和纪念馆馆长一起对国内韶山、井冈山、北京、天津、沈阳等地的革命题材纪念馆进行学习考察。行程中我们逐渐认识到，同类型纪念馆有强烈的共性，大部分都对称、端庄、肃穆。同时它们又都因题因地的不同而具有鲜明的个性。二者的结合，使这些馆各展风采，给人留下深刻的印象。每一座纪念馆在探讨建筑设计中，都重视功能，科学合理地安排功能布局、流线和各项设施，而能否突出建筑的纪念性，则是方案成败、优劣的关键之所在。

In December 2004 we undertook the design work and began feasibility study. For the project under the leadership of Yan'an and together with the memorial president we made a study tour on revolutionary memorial halls to Shaoshan, Jinggangshan, Beijing, Tianjin and Shenyang. In the process we gradually realized the memorial buildings of this kind have strong common features.

Most of them have a symmetrical layout with solemnity and dignity. At the same time they have their unique characteristics for their respective purposes and locations. The memorials left deep impression on us for their styles. In design of memorials importance should be attached on the functional layouts, circulations and installations. Priority of memorial features is the key for success of the design.

我们在延安革命纪念馆设计过程中体会到，对如此重大的纪念性建筑要突出其纪念性，必须建立起一个纪念性空间体系，统筹规划、建筑、内装、环艺各个要素，全方位地为突出主题而营造。为此，作了以下七方面的探索。

In design of such a large memorial priority should be given to its memorial characters and a memorial space system be created. Building layout, architectural art, interior design and environmental artistry are used to highlight the theme of the Hall. We have made efforts in the following 7 aspects.

利用山水格局烘托气势

延安位于黄河中游陕北黄土高原，全市丘陵山地占总面积的80.8%，川、塬占地19.2%。城市以宝塔山、清凉山、凤凰山为核心向三条川道延伸发展。延安革命纪念馆坐落在西北川中部的王家坪，西有枣园、杨家岭，这三处都是重要的革命旧址保护区。

Rivers and Hills are Used to Create Spacious Atmosphere

Yan'an is situated in the north Shaanxi yellow plateau in the middle reaches of the Yellow River. The city is occupied by hills by 80.8%. Flat low-lying land and tableland hold 19.2% of the city area.

纪念性空间体系的营造

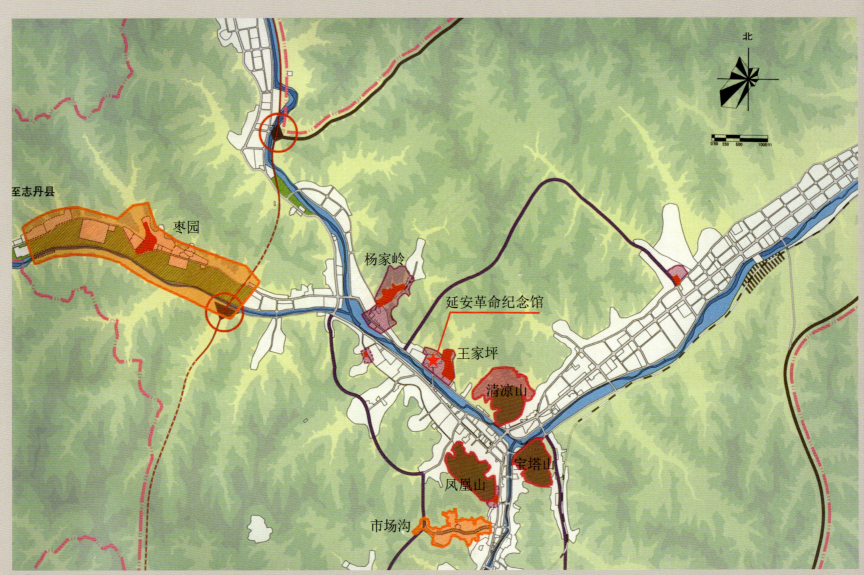

区位图 Location Map

Baota hill, Qingliang hill and Fenghuang hill stretch out from the centre area along the three river banks. The Yan'an Revolutionary Memorial Hall lies in Wangjiaping in the middle region of the northwest flat land with Zaoyuan and Yangjialing to its west. The three places are all important revolutionary sites for protection.

纪念馆坐北朝南，背靠赵家峁，东西两侧山势呈环抱状，面临延河。本次设计尊重纪念馆与斜跨延河的彩虹桥所形成的南北轴线，以此作为新馆轴线。为适应新馆建设的需要将馆址基地从原10公顷扩大至15.78公顷。基地沿延河展开宽度约500米，从沿河道路至赵家峁山麓深约260米不等。南北中轴从路边达山麓336米。这一用地格局的确定使这片背山面水的坪地为纪念馆的合理布局提供了颇有气势的空间环境。

纪念性内容，丰富人性化的服务手段。这样既增加了绿地面积，提高了环境品位，又可使赵家峁南坡的松柏区与馆区绿地上下呼应，融为一体。由此纵观自南而北由彩虹桥、纪念馆大门、纪念广场、旱喷水池、毛主席雕像、纪念馆主入口、序厅、纪念园景点、赵家峁松柏区构成了丰富多彩的革命纪念景观轴，从而形成纪念性空间体系的脊梁。

The Memorial faces south with Zhaojiamao in the rear. It is enclosed by hills to the east and west. The Yanhe River flows in the front. The design respects the north-south axis formed by the Hall and the Caihong Bridge which spans the Yanhe River. With this axis line the new hall area is enlarged to 15.78 hectares from 10 hectares of the old one. The site is 500 m wide along the Yanhe River and 260 m deep or so from Yanhe roads along the river to Zhaojiamao hill foot. The north-south axis line is 336 m long from the road side to the hill foot. The allocated site with hills in the rear and a river in the front provides a reasonable and elegant space environment.

广场建筑园区融为一体

在总体布局上，以彩虹桥为导向，沿轴线自南而北布置了纪念广场、纪念馆建筑、纪念馆园区三大部分。三者有机结合、穿插有序、融为一体。为适应延安市举行大型群众性活动的要求，广场设计为29000平方米。毛主席铜像居于广场中部。地面铺装设计有同心圆的大型弧线节理，其上布有草皮、花池，有节奏地为宽阔的广场增加了向心的凝聚感。纪念馆建筑背山面河，中规中矩的"冂"形布局与广场有机结合，形成吸纳之势。在纪念馆周边至山麓规划设计了革命园区。园内通过以"胜利之路"红飘带的形象组织游览路线，沿线设置纪念性景点和服务设施，以拓展革命

The Memorial is Integrated with the Square and the Park

In the general layout the Caihong Bridge forms a guide place. The three parts of the memorial square, the memorial building and memorial park are arranged along the axis line from south to north. They are combined organically in integration in an orderly sequence. To meet requirements of large scale mass activities in Yan'an a square of 29000 m^2 is designed with the bronze statue to Chairman Mao Zedong standing in the middle. The ground is finished rhythmically with lawns and flower beds in large circular shapes. The circles with one same centre add cohesion sense toward the centre. The Memorial has a neat and tidy layout with a river in the front and a hill in the rear. In combination with the square, the "冂"type layout appears to be appealing. The surrounding area from the Hall to the hill foot is planned as a revolutionary park zone. In the zone the "victory road" in red ribbon shape is the way for visitors. Along the way there are some memorial spots and service facilities to expand memorial contents and service means. In doing so the green area is increased and the environmental quality improved, and the pine and cypress zone on the south hill of Zhaojiamao and park area echo each other to form an integral whole. Along the axis line from south to north are the Caihong Bridge, gate of the Memorial Hall, memorial

纪念性空间体系的营造

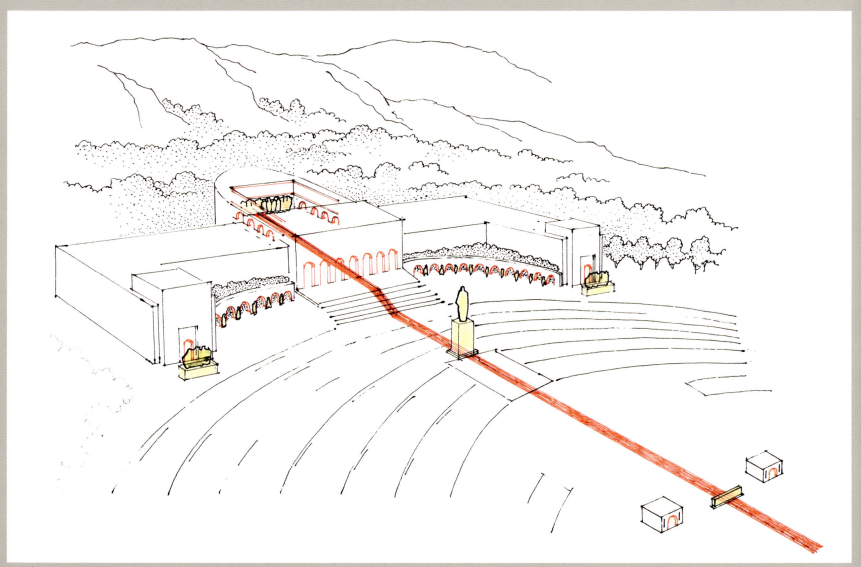

纪念性空间体系景观轴 Vision Axis of Memorial Space System

square, dry fountain, statue to Chairman Mao, main entrance to the Hall, prelude hall, park scenery spots and pine and cypress zone of Zhaojiamao. All these form a rich and colourful landscaping axis and a spine of memorial and revolutionary spaces.

建筑体型设计超常向量

纪念馆采取横向展开的布局，建筑体型突出中部门廊体块的高度，两翼平直延伸，东西两端再稍作凸起，形成稳重的"山"形立体轮廓，也成为广场上毛主席铜像的坚实背景。考虑到纪念馆东、西、南三面城镇化发展已经高楼林立，仅三层高的纪念馆形象很难突出。根据建设规模与功能布局，将纪念馆设计为东西长222米，南北深78.5米的"⊓"形。

Building Shape is Designed with Abnormal Vector

The Memorial spreads out transversely. The building has a very high porch with its two flanks extruding horizontally and the east and west ends turning up to form a "山" type building shape, and it provides a background for the bronze statue to Chairman Mao. In consideration of the tall buildings already erected to the east, west and south of the Hall the image of the 3-storey Hall building is difficult to be outstanding. In accordance with the building scale and function layout the Hall is designed in "⊓" shape with a length of 222 m from east to west and a depth of 78.5 m from north to south.

在建筑横向水平尺度上"一压群芳"，其超常的向量所体现的张力和呈围合态势的控制力，奠定了纪念馆在西北川乃至于延安市区中实现纪念性的基础。

The transverse scale of the building is dominating. The extraordinary length appears to have expansion and containing power. This provides a base for the memorial features of the building in the northwest flat land and even so in Yan'an urban area.

延安建筑文脉继承发扬

建筑在黄土高原的窑洞建筑朴实无华而又无比坚毅。它对于延安早已超出了一般地方建筑形式的意义。革命战争时期，延安上自中共中央领导，下至各级干部、军民百姓都是以窑洞为家。到今天，枣园、杨家岭、王家坪革命旧址中的窑洞与那些因山就势的孔孔窑洞民居，都有了延安革命精神的象征意义。因此，"窑洞"的造型就理所当然地成为纪念馆建筑艺术的母题。在革命战争中延安也曾出现少数新建的公共建筑，如杨家岭"七大"会堂、中央办公厅就是最突出的代表作，地方材料、简洁体型、竖长的条窗与洞窗和竖线条的立面处理，挺拔有力、别具一格。纪念馆室内外设计多处运用"窑洞"母题和"七大"会堂等建筑元素，引发了人们对革命历史的缅怀和无限遐想。

Yan'an Architectural Context is Inherited and Developed

The cave dwellings dug out of earthen hills on the Loess Plateau of northwest China are plain and firm. In Yan'an they mean more than an ordinary building form. During the period of the revolutionary war all the cadres of different ranks, soldiers and civilians in Yan'an took caves as their dwellings. Today the caves in Zaoyuan, Yangjialing and Wangjiaping revolutionary sites and those cave dwellings on the hills have become symbols of Yan'an revolutionary spirit. Therefore the shape of cave dwellings is taken as the architectural art motif of the Memorial. In the time of the revolutionary war some public buildings of new type were constructed, of which the hall of the 7th representatives conference of the Chinese Communist Party in Yangjialing and the general office of the Central Committee are typical. Local materials, simple shapes, strip windows and cave windows are treated in a special way to make them to be tall and straight.

Repeated use of the cave motif and building elements of the hall of the 7th representatives conference make people to recall the revolutionary history and give wings to imagination.

枣园山村 Zaoyuan Village

中央大礼堂 Central Auditorium of the Cpc Central Committee

中共中央办公厅远眺 Full View of General Office of the Cpc Central Committee

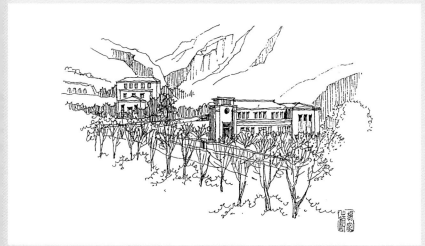
杨家岭 Yangjialing

枣园旧居 Zaoyuan Former Residence

中共中央办公厅 General Office of the Cpc Central Committee

毛主席旧居 Former Residence of Chairman Mao

建筑主体形象蕴含喻意

纪念馆主入口门廊设在高于广场3.9米的首层中央。门廊为全馆最高体部，七开间窑拱式券廊，以暗合在延安举行的历史性中共第七次代表大会。券廊上方花岗石墙面正中为阳刻党徽，其下有"延安革命纪念馆"七个阴刻金字。通向门廊的大石阶平缓、宽阔，全高分为三台，以隐喻中国共产党在延安经历了土地革命、抗日战争、解放战争三个阶段。台阶东西两侧设有系列大型纪念灯柱，以示革命事业之辉煌。整组入口设计主体突出，导向明确、气势恢宏，使建筑与广场相互映衬。

The Shape of the Main Building Implies Meaning

The main entrance porch of the Memorial is in the middle of the ground floor, whose level is 3.9 m higher than the square. The porch is the highest part of the building and has 7 cave arched spans that symbolizes the 7th representatives conference of the Chinese Communist Party. The party campaign emblem is carved in relief in the middle of the granite wall on the upper part of the porch. Under the emblem are the seven golden Chinese words "延安革命纪念馆" (Yan'an Revolution Monument) in intaglio. The grand stone steps to the porch are open and flat and has three landings that implies the three stages which the Chinese Communist Party experienced in Yan'an i. e. land revolution, anti-Japanese war and liberation war. On both sides of the steps are large lamp posts indicating the glory of the revolutionary cause. The Hall is designed in grandeur to make the main building prominent and the guide way clear. The magnificent building and the square set each other off.

纪念性雕塑突出主题

建筑只能抽象地表达思想，而雕塑则能具体地表现主题。纪念馆在室内和室外都设计了有分量的大型群雕。室内序言大厅中是"党中央领导和人民在一起"。室外广场上三个层次的雕塑是"从胜利走向胜利"。在广场中心部位，以旱喷水池衬托着总高16米的毛泽东主席铜像，以象征毛泽东思想指引中国革命从胜利走向胜利；在纪念馆主入口两侧弧形窑洞式纪念墙衬托之下，18尊工、农、兵、学、商的雕像环毛主席像而立，表现了革命战争年代边区军民对毛主席和中国共产党的拥护与爱戴；在纪念馆东西两个次入口前的平台上设计有"延安精神放光芒"的群雕。这些大型雕塑的设置丰富了广场的文化内涵，深入了革命精神，突显了整个环境的纪念性。

Memorial Sculptures Highlight the Theme

Architecture can only express ideas in abstraction while sculptures can definitely show the theme. Large group sculptures are designed inside and outside the Hall. In the prelude hall is the sculpture of "the Party leaders and the people in the same place". On the outside square are three group sculptures of "from victory to victory". In the centre of the square is the dry fountain that sets off the 16 m bronze statue to Chairman Mao Zedong symbolizing the Chinese people marching forward from victory to victory under the guidance of Mao Zedong's thought. In front of the cave type memorial walls on both sides of the main entrance are 18 sculptures of workers, peasants, soldiers, students and merchants standing around the bronze statue of Chairman Mao displaying support and adoration of the army and people in the border region during the revolutionary war time. On the terraces in the two side entrances are group sculptures of "glorious Yan'an spirit". These large group sculptures on the square add cultural connotation, express revolutionary spirit and emphasize memorial characteristics of the environment as a whole.

序厅艺术空间精神感召

在纪念馆门厅正前方，设计了24米×29米两层通高的序厅，以期通过空间内的纪念性艺术效果更好地发挥革命精神的感召力和教育作用。序厅正位是延安时期五大书记为中心的大型群雕，主题是"党中央和人民在一起"。群雕背景是以宝塔山为主景的延安全景浮雕，东侧为"黄河壶口瀑布飞流"，西侧为"黄帝陵古柏参天"。三幅浮雕连成一体气象非凡。序厅两侧浮雕以下为窑洞形的连续拱券。序厅屋顶是整片轻钢索结构的玻璃天窗，使蓝天白云与纪念雕像同在，明媚阳光与拱券窑洞相映。厅内一片光明，表现了延安革命年代朝气蓬勃的精神。这里应是回顾革命历史、缅怀革命先贤的革命圣殿。

The Art Space of the Prelude Hall Presents Spiritual Qualities

In front of the entrance lobby a 24 m x 29 m prelude hall of a 2-floor height is designed. The memorial art effects in the hall will better display inspiring and educational power. In the centre is the large group sculpture of "the Party leaders and the people in the same place" with the five figures of the secretaries in the Yan'an period standing in the middle. The panoramic relief sculpture has the Baota Hill as the main feature and provides a background for the group sculpture. To the east is the sculpture of "the flying Yellow Hukou Waterfall" and to the west is that of "the Yellow Emperor Mausoleum with ancient cypress reaching into the sky". The three relief sculptures present a grand air. Under the sculptures are continuous arches in cave shapes. The roof of the prelude hall is constructed with light weight steel cables and glass skylights. The memorial sculptures stand under the blue sky and white clouds. The cave like arches appear in the bright sunlight. All this expresses that the Yan'an revolutionary spirit is full of vigour and vitality. This place will play as a revolutionary temple for people to recall the revolutionary history and cherish the memory of former revolutionary worthies.

经过四度酷暑严寒的奋战，2009纪念馆竣工。建筑、总图、内装、环艺各方从方案阶段就开始切磋磨合，通力合作，终于实现了设计目标，受到各界广泛好评，重建的延安革命纪念馆于2009年被评为"中华人民共和国成立六十周年百项经典工程"。

After 4-year endeavours the Memorial was completed in 2009. Through collaboration between architecture, general planning, interior design and environmental skill the goals of the design were realized and the Hall has received good remarks from various circles. The Yan'an Revolution Monument Hall was appraised as "a classic project in the 60 anniversary of the People's Republic of China".

纪念性空间体系的营造

| 名称：延安革命纪念馆 | Project: Yan'an Revolution Monument |

名称：延安革命纪念馆
位置：陕西省延安市
占地：15.87 hm²
建筑面积：29853 m²
建设单位：延安革命纪念馆
设计单位：中国建筑西北设计研究院
施工单位：陕西建工集团总公司
设计时间：2004年12月~2006年10月
竣工时间：2009年8月
方案设计人：张锦秋
项目负责人：张锦秋　王　军
主要设计人：
　建筑：张昱旻　徐　嵘　张小茹
　室内：丁　梅　李午亭
　总图：陈初聚
　结构：韦孙印　王洪臣
　给排水：张　军　秦发强
　暖通：殷元生　薛　洁
　电气：杜　乐　曹维娜
　景观：李　毅　曾　健
主要技术经济指标：
　占地面积：15.87hm²（含赵家峁部分山地）
　总建筑面积：29853 m²
　建筑密度：9%
　容积率：0.19
　绿地面积：83625 m²
　绿地率：60.7%
　广场面积：28977 m²
　室外停车场面积：2300 m²
　停车位126辆：
　　室外 48辆（大）
　　库内 26辆（大），52辆（小）
　陈列面积：12000 m²

Project: Yan'an Revolution Monument
Location: Yan'an city, Shaanxi Province
Site area: 15.87 hm²
Floor area: 29853 m²
The client: Yan'an Revolution Monument
Design unit: China Architecture Northwest Design and Research Institute
Contractor: Shaanxi Construction Group Co.
Design time: December 2004 ~ October 2006
Completion: August 2009
Scheme designer: Zhang Jinqiu
Project leader: Zhang Jinqiu　Wang Jun
Designers:
　Architecture: Zhang Yumin, Xu Rong and Zhang Xiaoru
　Interior: Ding Mei, Li Wuting
　General plan: Chen Chuju
　Structure: Wei Sunyin, Wang Hongchen
　Water supply & drainage: Zhang Jun, Qin Faqiang
　Heating ventilation & air conditioning: Yin Yuansheng, Xue Jie
　Electrical: Du Le, Cao Weina
　Landscaping: Li Yi, Zeng Jian
Main technical data:
Site area: 15.87 hm² (Including some hill area of Zhao jiamao)
Total floor area: 29853 m²
Building density: 9%
Floor area ratio: 0.19
Green area: 83625 m²
Percentage of green area: 60.7%
Square area: 28977 m²
Outdoor parking area: 2300 m²
Parking capacity:
　Outdoor 48 (Large Cars)
　Garage: 26 (Large Cars), 52 (Small Cars)
Exhibition area: 12000 m²

001
馆区鸟瞰
Bird's-eye View

002

003

002
纪念馆正面透视
Front Perspective Drawing of the Hall

003
纪念馆背面透视
Back Perspective Drawing of the Hall

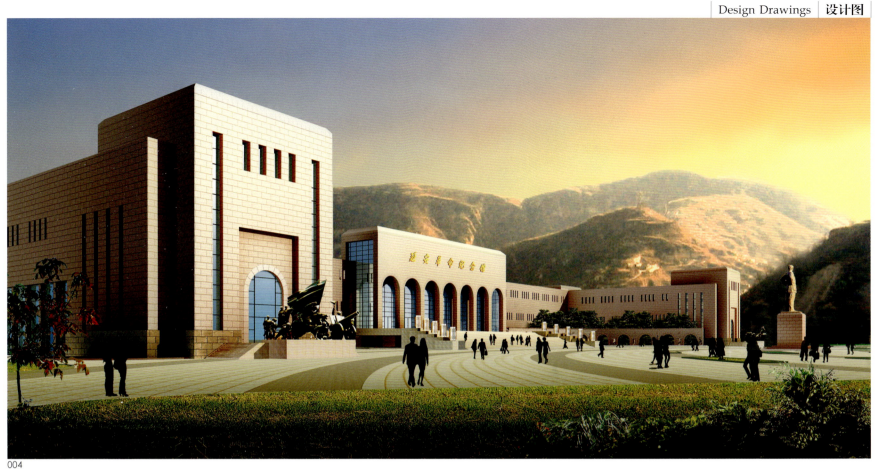

004

005

006

004	005	006
纪念馆西南侧透视	纪念馆入口透视	纪念馆纪念墙透视
Southwest Perspective Drawing of the Hall	Entrance Perspective Drawing of the Hall	Memorial Wall Perspective Drawing of the Hall

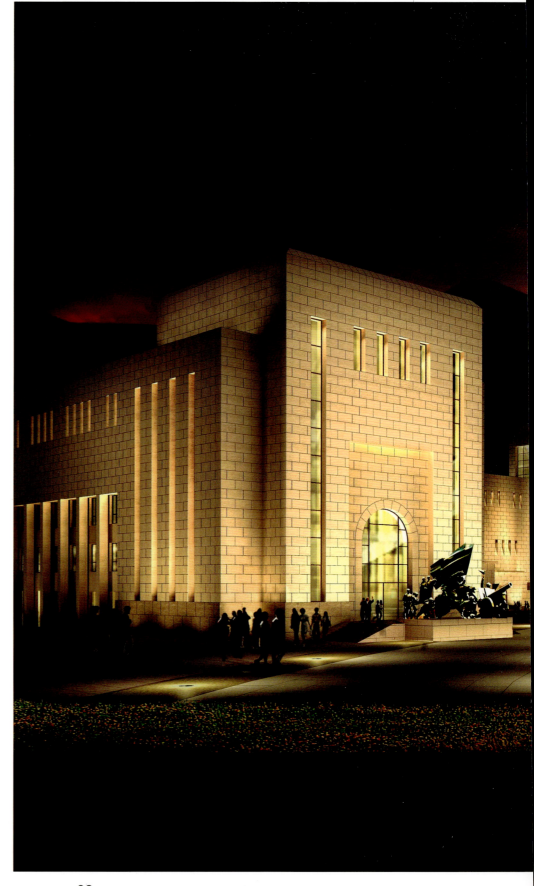

007
纪念馆夜景
Night View of the Hall

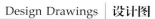

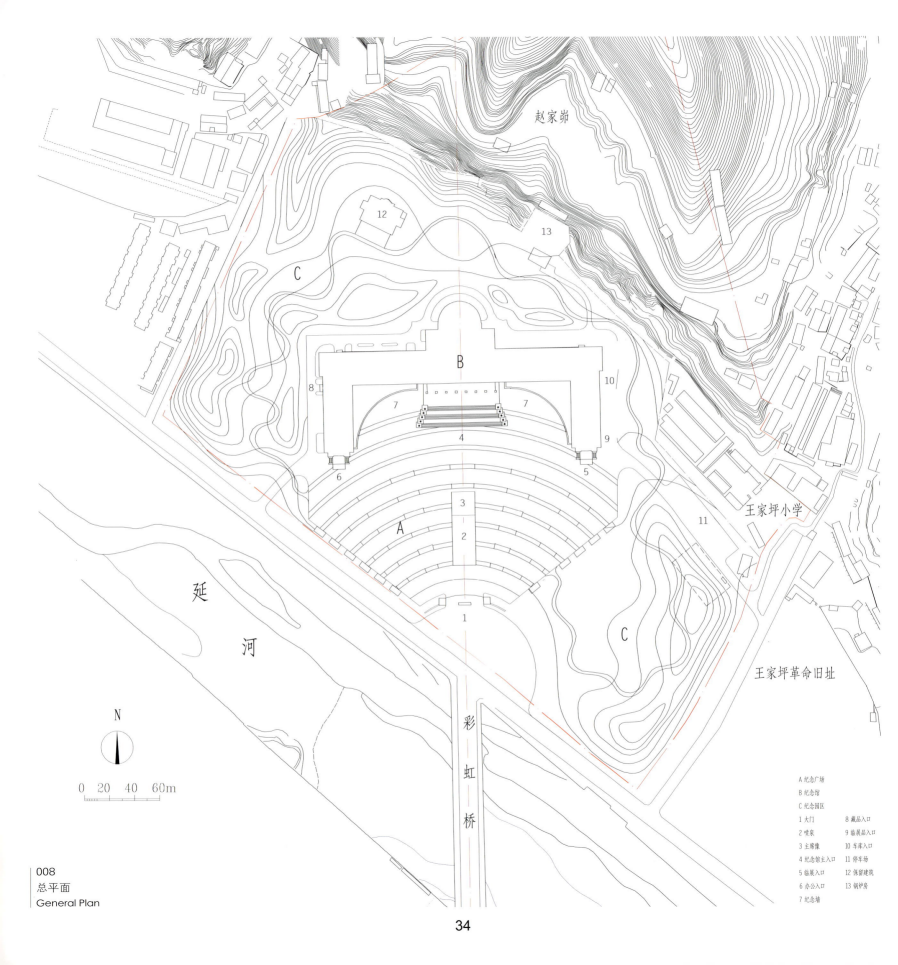

008
总平面
General Plan

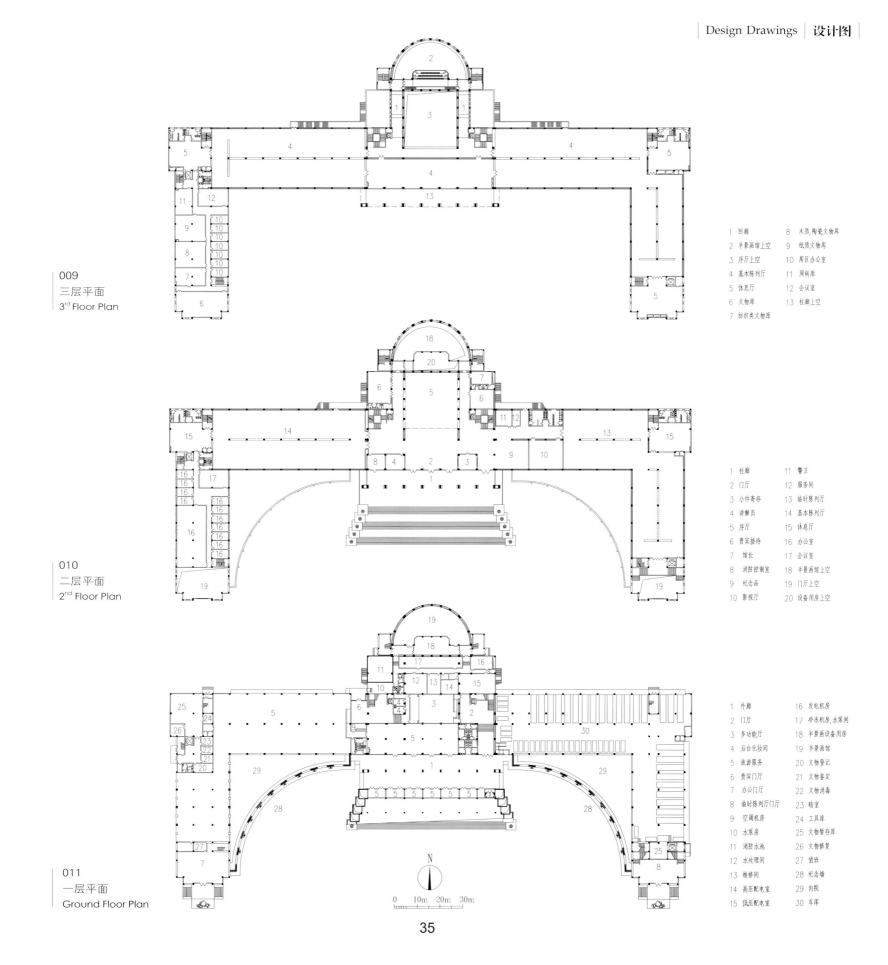

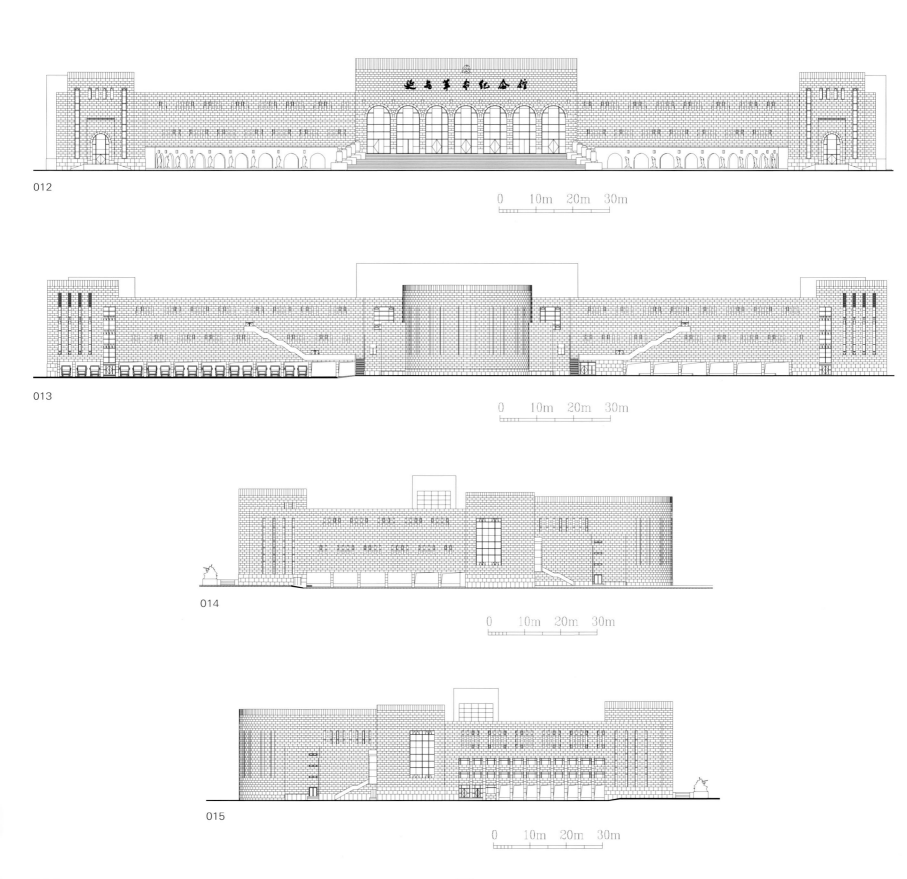

| Design Drawings | 设计图

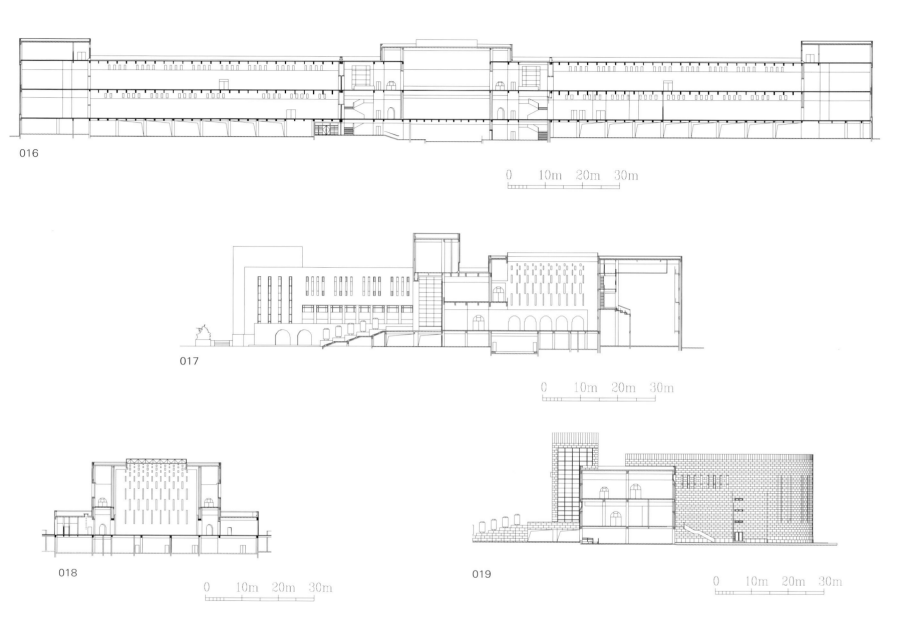

012	013	014	015	016	017	018	019
南立面图	北立面图	东立面图	西立面图	东西总剖图	南剖面图	序厅剖图	展厅南北剖图
South Elevation	North Elevation	East Elevation	West Elevation	West-East General Section	South Section	Section of the Prelude Hall	North-South Section of the Exhibition Hall

| Environment | 大环境 |

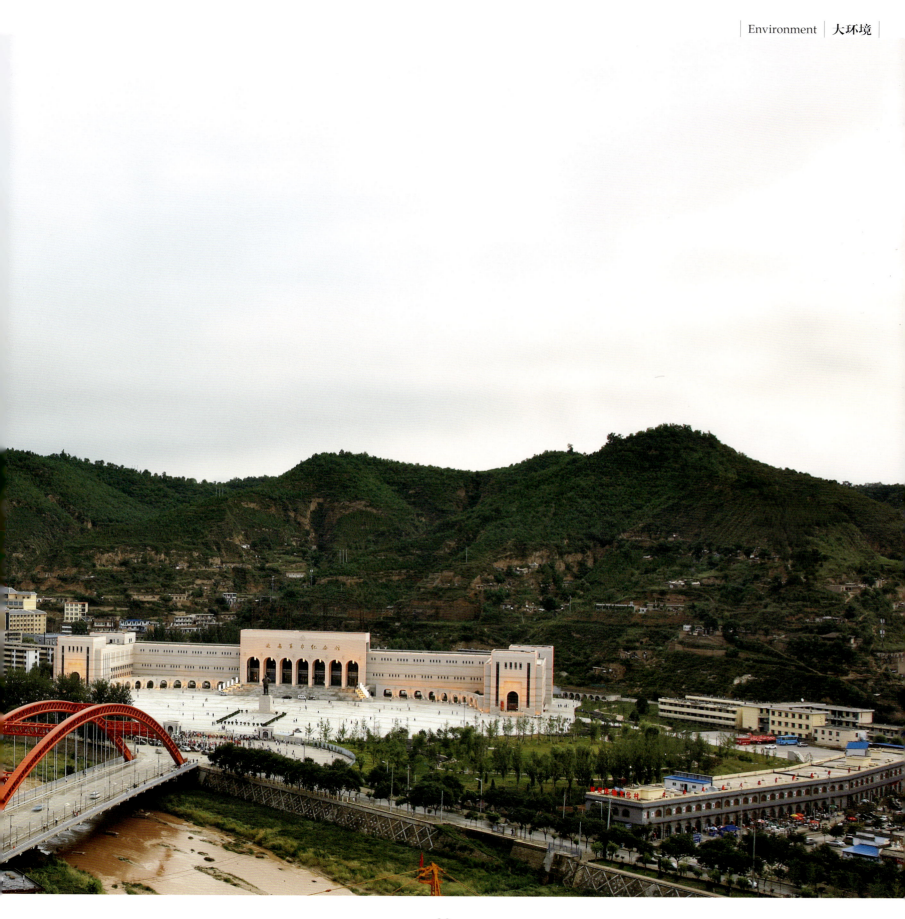

020
西北川西部全景
West View of the Northwest Flat Land

021
纪念馆原状鸟瞰
Bird's-eye View of the Old Memorial

022
西北川东部全景
East View of the Northwest Flat Land

023
纪念馆远景平视
Distant View of the Memorial

024
从东侧鸟瞰纪念馆
Bird's-eye View of the Memorial

025
从东南鸟瞰纪念馆
Bird's-eye View of the Memorial from Southeast

Environment | 大环境

Environment | 大环境

027

028

026 △
从西南鸟瞰纪念馆
Bird's-eye View of the Memorial from Southwest

027 △
毛主席像眺望
Distant View of Chairman Mao Statue

028 △
从东北鸟瞰广场
Bird's-eye View of the Square from Northeast

029

030

029
西北川川道夜景东望
Night View of theNorthwest Flat Land from East

030
纪念馆夜景鸟瞰
Bird's-eye Night View of the Memorial

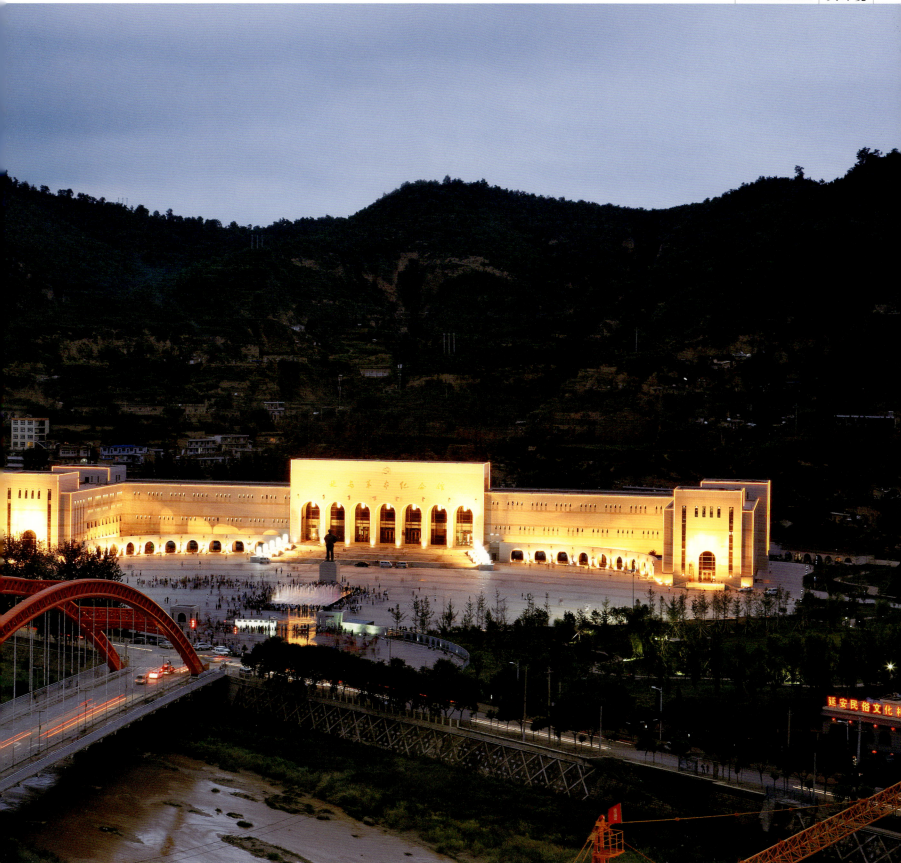

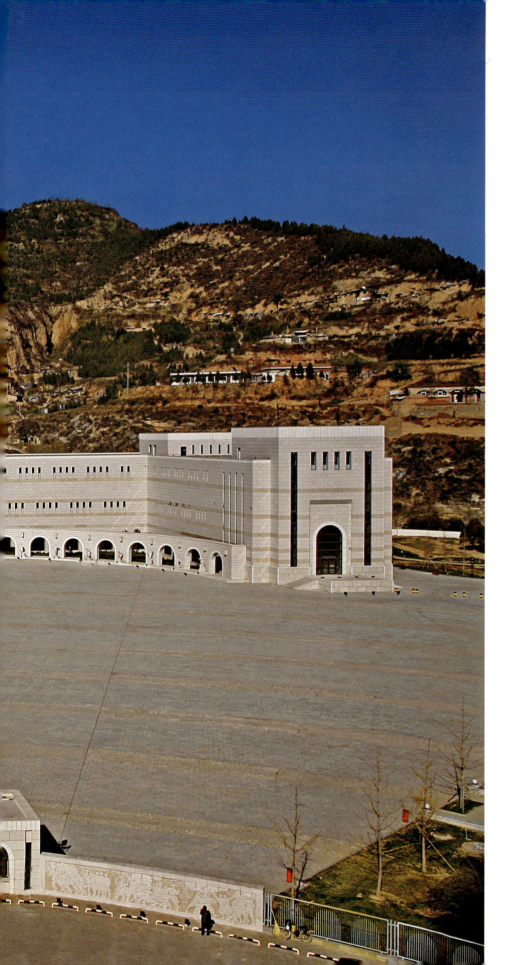

Building Shape Space | 建筑形体空间

031
纪念馆正面全景
Front Full View of the Memorial

| Building Shape Space | 建筑形体空间

032
从东南侧绿地中看全景
Full View from Southeast Green Area

Building Shape Space | 建筑形体空间

033
正面全景
Front Full View

Building Shape Space | 建筑形体空间

034
西南侧面全景
Southwest Full View

035
"⊓"内从东俯瞰
Bird's-eye View of "⊓" Inside from East

036
东南侧全景
Full View from Southeast

Building Shape Space | 建筑形体空间

Building Shape Space | 建筑形体空间

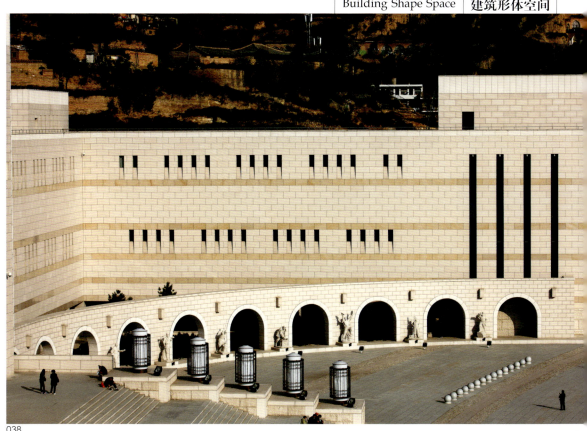

038

037
" ⊓ " 内东侧全景
East Full View of " ⊓ " inside

038
" ⊓ " 内东侧全景
East Full View of " ⊓ " inside

Building Shape Space | 建筑形体空间

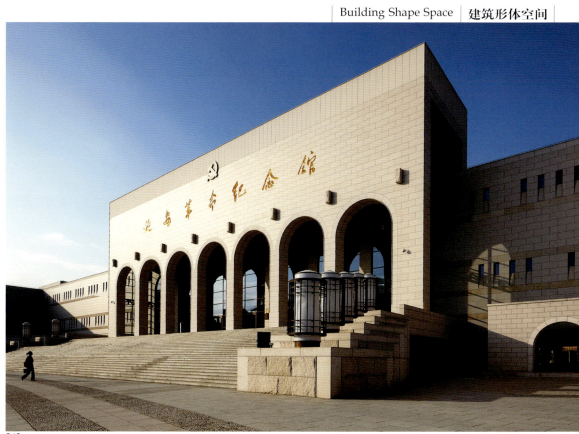
040

041

043

044

039
东纪念墙与建筑全景
East Memorial Wall and Full View of the Building

040
从东南看中部
Central View from Southeast

041
东纪念墙俯视
Bird's-eye View of the East Memorial Wall

042
从西南近瞰广场
Close bird's-eye View of the Square from Southwest

043
门廊前东望
East View from Porch

044
门廊前西望
West View from Porch

045
"冂"内东侧夜景
East Night View of "冂" inside

Building Shape Space | 建筑形体空间

046

047

046
门廊透视
Porch Perspective

047
门廊三开间
3-span Porch

048
门廊与台阶、灯
Porch, Steps and Lamps

Main Building · Porch Details | 建筑主体形象·门廊特写

049
门廊内景
Inside View of the Porch

050
门廊内景
Inside View of the Porch

Main Building · Porch Details | 建筑主体形象·门廊特写

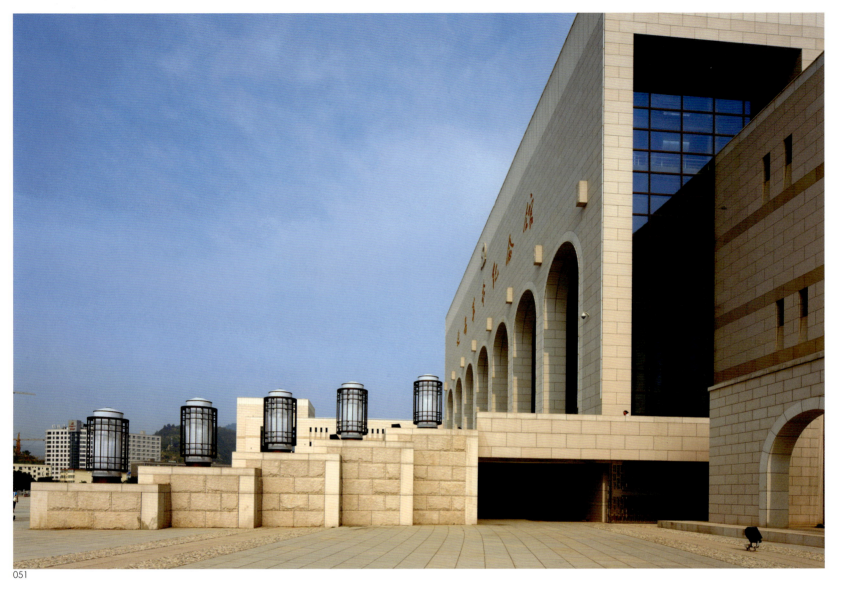

051

052

053

Main Building · Grand Steps Details | 建筑主体形象·大台阶特写

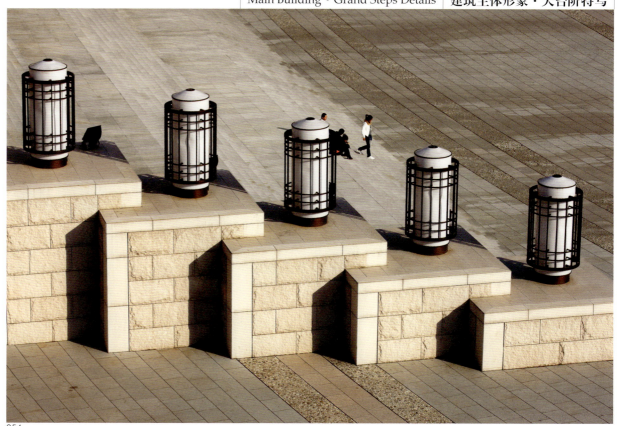

054

051
大台阶、架空平台外侧
Grand Steps and Elevated Terrace outside

052
大台阶与灯柱
Grand Steps and Lightposts

053
大台阶俯视
Bird's-eye View of the Grand Steps

054
大台阶与灯柱
Grand Steps and Lightposts

055
大台阶与灯柱
Grand Steps and Lightposts

055

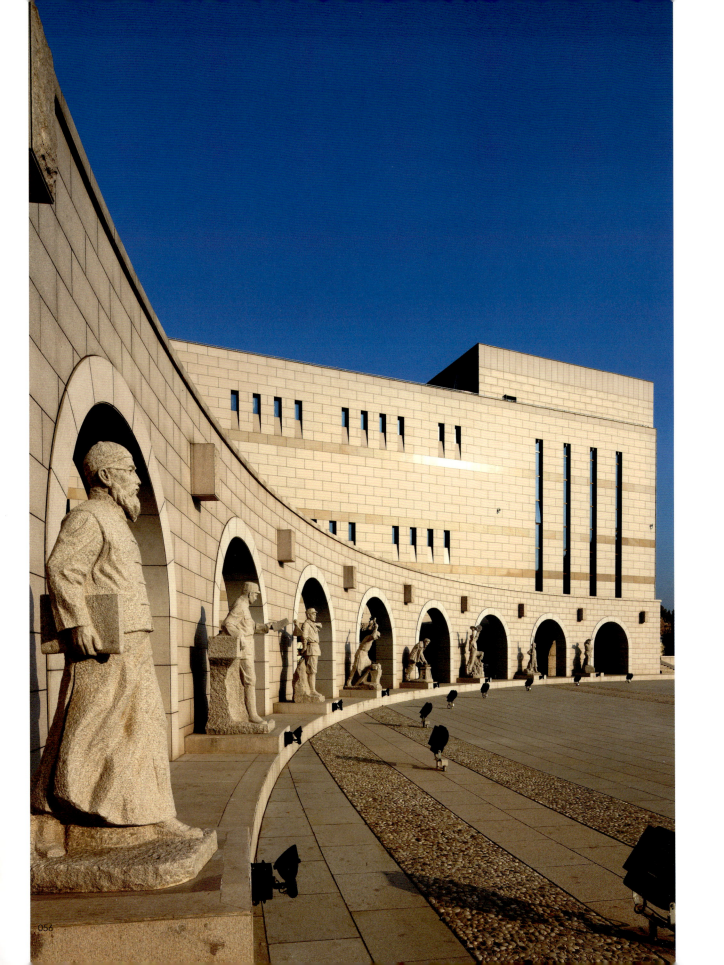

056
东纪念墙
East Memorial Wall

057
东纪念墙
East Memorial Wall

Main Building · Memorial Wall Details | 建筑主体形象·纪念墙特写

058
一孔窑洞
Cave Dwelling

059 △
从纪念墙内外望
Outside View through the Memorial Wall

060
园区围栅
Park Fence

Main Building · Memorial Wall Details 建筑主体形象·纪念墙特写

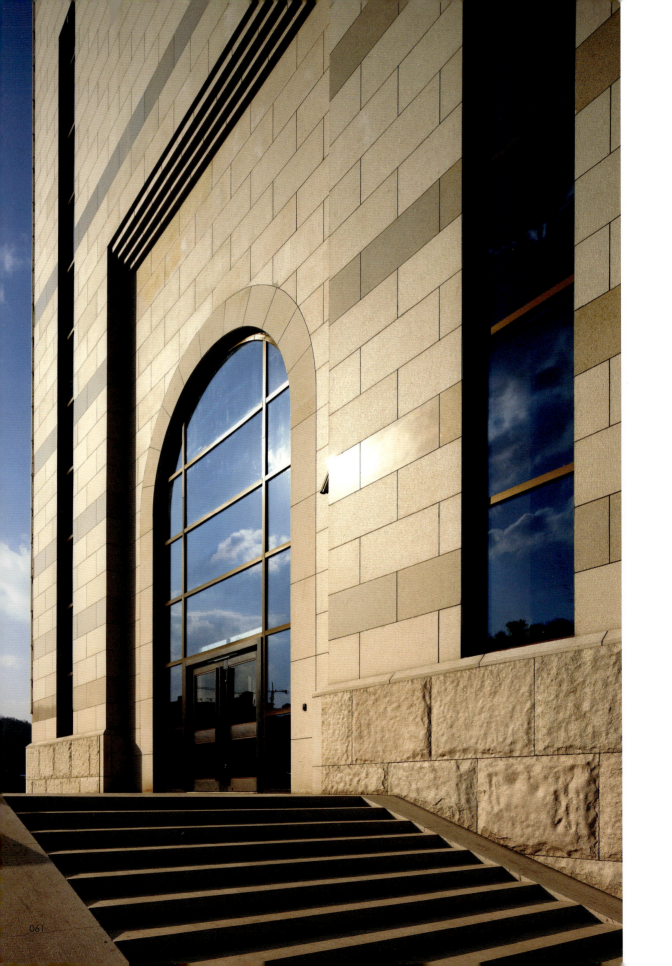

061
办公楼入口
Office Building Entrance

062
办公楼东南视角
Office Building from Southeast

Building Elevation Detail | 建筑立面特写

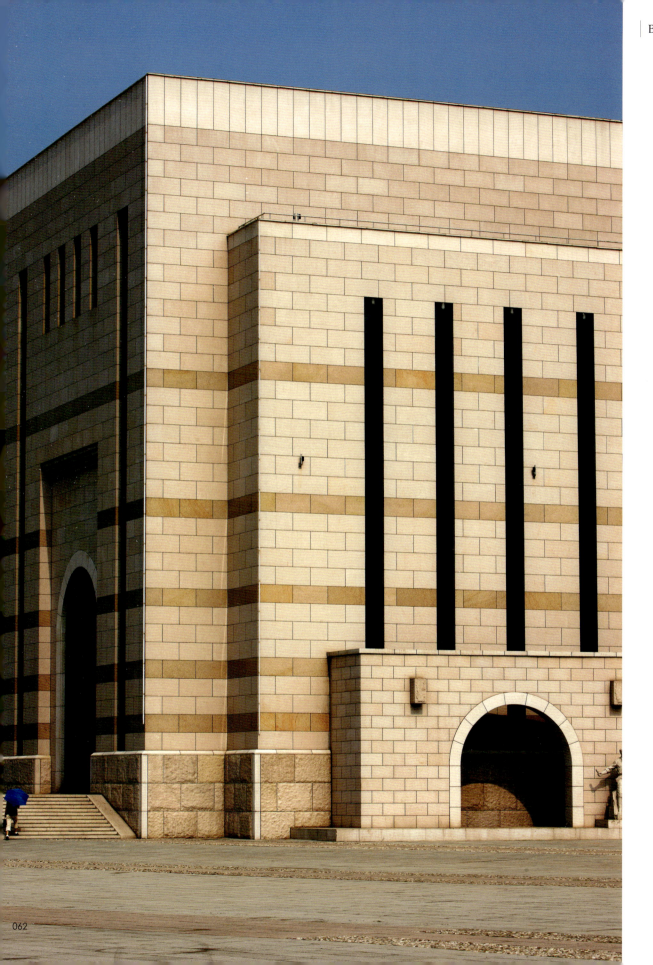

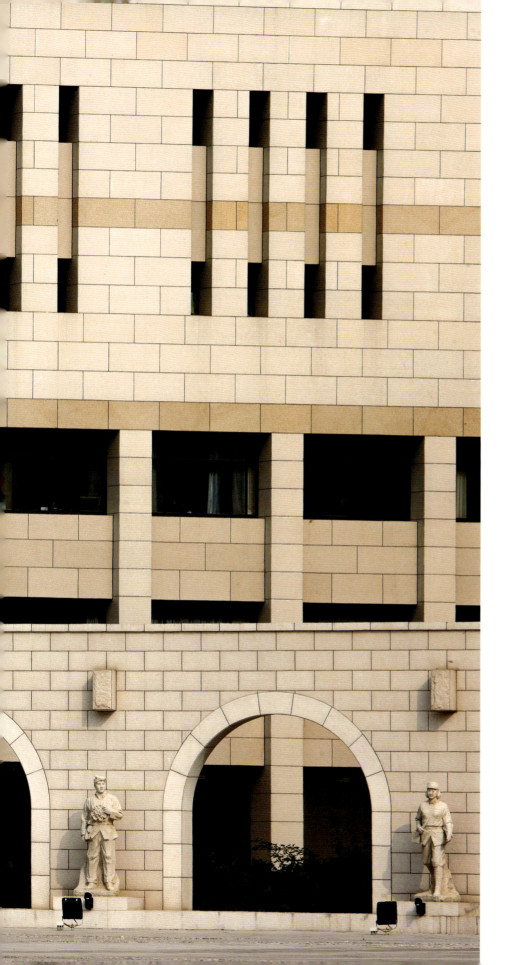

Building Elevation Detail | 建筑立面特写

063
办公楼东墙与纪念墙
East Wall of the Office Building and the Memorial Wall

064
办公楼西墙
West Wall of the Office Building

065
展厅外墙窗户特写
Details of the Windows in the External Walls of Exhibition Hall

| Building Elevation Detail | 建筑立面特写

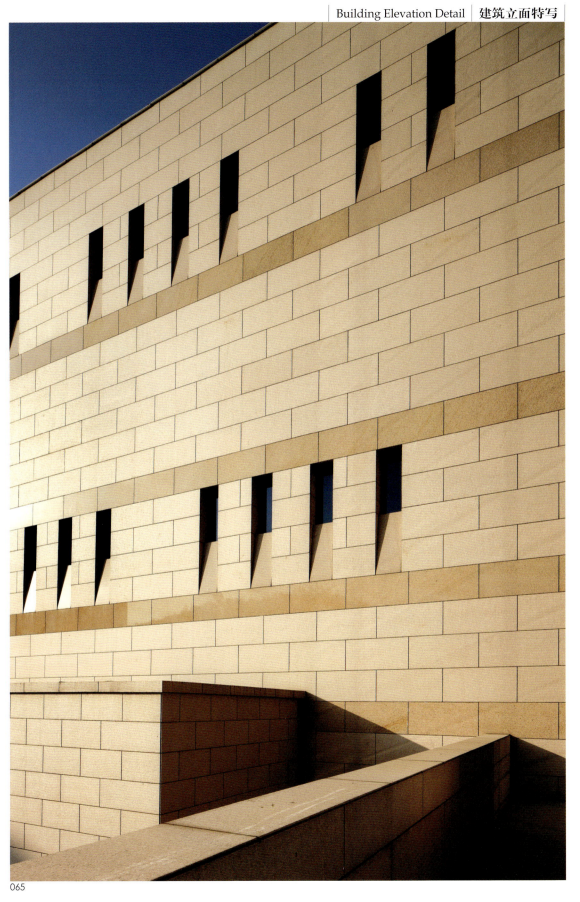

066
东翼东墙
East Wall of the East Wing

067
中轴区与东翼北墙
Central Axis Zone and North Wall of the East Wing

068
环景画馆外观
External View of the Circular Gallery

Building Elevation Detail | 建筑立面特写

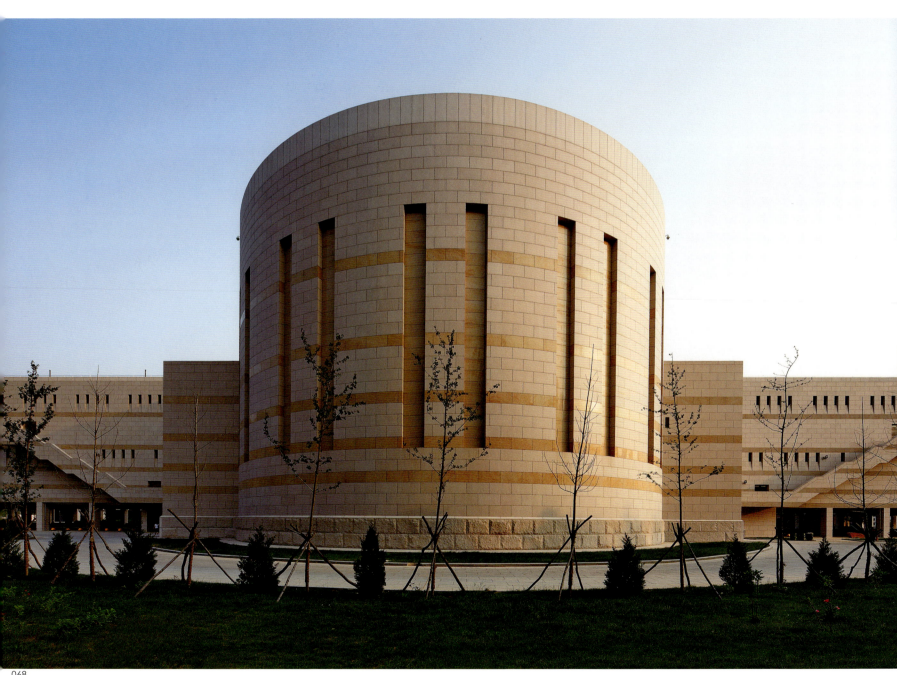

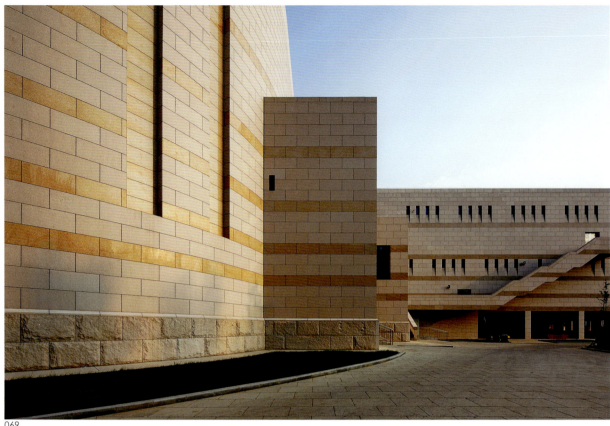

069

070

069
中轴区与西翼北墙
Central Axis Zone and North Wall of the West Wing

070
西翼北墙
North Wall of the West Wing

071
架空休闲厅与纪念墙间绿地
Green Area between the Elevated Leisure Hall and the Memorial Walls

072
序厅全景
Full View of the Prelude Hall

| Building Elevation Detail | 建筑立面特写

Interiors · Prelude Hall | 室内·序厅

073

Interiors · Prelude Hall | 室内·序厅

073
序厅全景
Full View of the Prelude Hall

074
序厅雕塑全景
Full View of the Sculptures in the Prelude Hall

075

076

| 075 | 076 | 077 | Interiors · Prelude Hall 室内·序厅 |

075 序厅透视
Perspective of the Prelude Hall

076 序厅西面正视
West View of the Prelude Hall

077 序厅南面正视
South View of the Prelude Hall

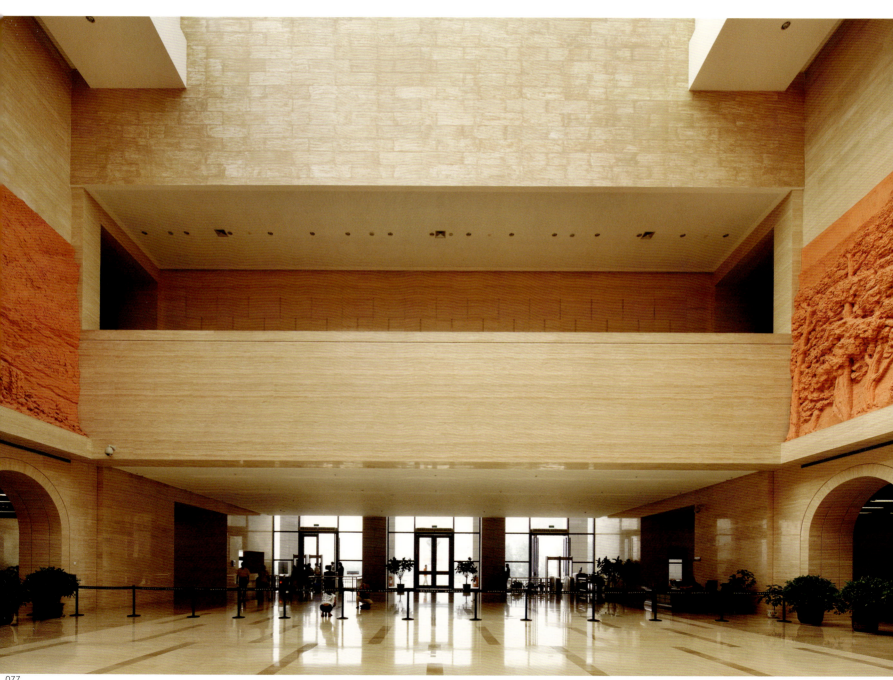

077

Interiors · Circulation Space | 室内·交通空间

078
序厅东廊
East Corridor of the Prelude Hall

079
序厅西廊
West Corridor of the Prelude Hall

080
从序厅东廊看贵宾休息入口
VIP Lounge Viewed from the East Corridor of the Prelude Hall

081
从序厅东廊看序厅
Prelude Hall Viewed from the East Corridor of the Prelude Hall

082
序厅东廊与楼、电梯厅
East Corridor of the Prelude Hall and the Staircase and Lift Hall

Interiors · Circulation Space | 室内·交通空间

081

082

083

084

083
东交通厅空间
East Circulation Hall Space

084
楼梯口与纪念品店
Stairs and Souvenirs Shop

085
楼梯间
Staircase

086
楼梯间
Staircase

| Interiors · Circulation Space | 室内·交通空间 |

085

086

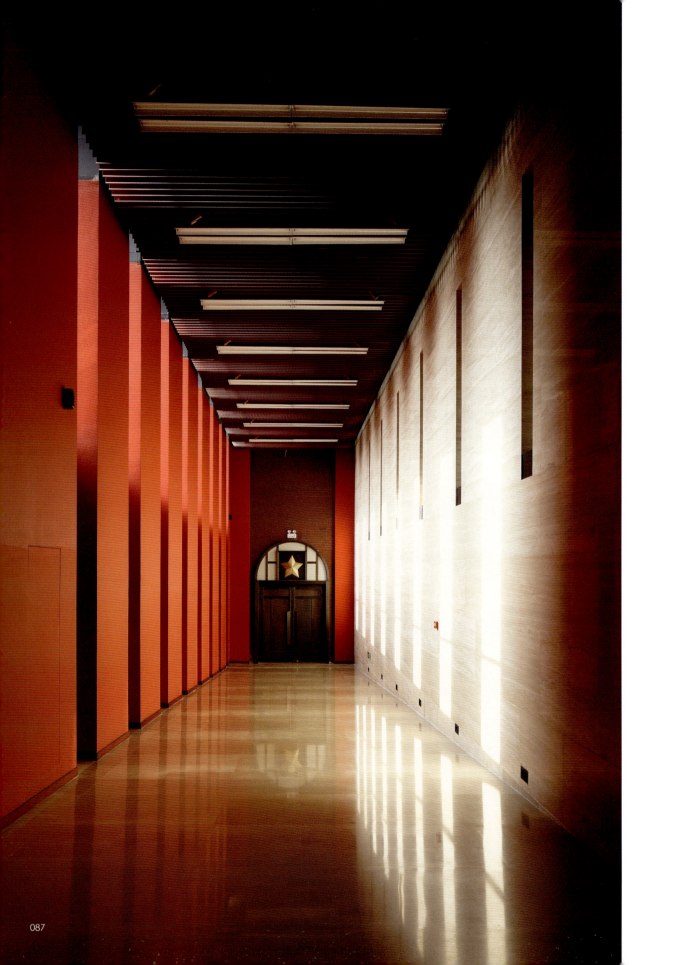

087
半景画厅候场空间
Waiting Space of the Semicircle Painted Hall

088
办公楼门厅
Entrance Hall of the Office Building

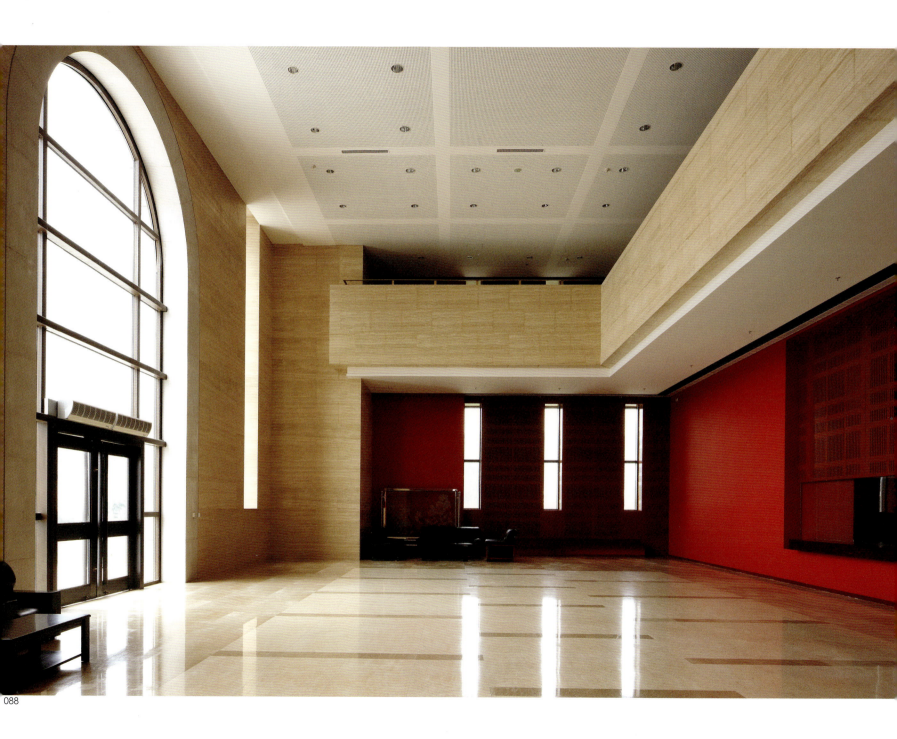

090

091

089
展厅
Exhibition Hall

090
展厅
Exhibition Hall

091
展厅
Exhibition Hall

092

093

092
展厅
Exhibition Hall

093
展厅
Exhibition Hall

094
展厅
Exhibition Hall

Interiors · Exhibition Hall Space | 室内·展厅空间

095
展区观众休息厅
Exhibition Region
Visitors Lounge

096
展区观众休息厅
ExhibitionRegion Visitors Lounge

097
影视厅
Video Hall

Interiors · Public Activities Space | 室内·公共活动空间

098
报告厅
Lecture Hall

Interiors · Public Activities Space | 室内·公共活动空间

099
底层报告厅外休息厅
Visitors Lounge of the Lecture Hall on the First Floor

100
底层报告厅外休息厅
Visitors Lounge of the Lecture Hall on the First Floor

101
底层门厅
Entrance Hall of the First Floor

Interiors · Public Activities Space | 室内·公共活动空间

102
西翼底层架空休息茶饮厅
Elevated Leisure and Tea Lounge on the First Floor of the West Wing

全国爱国主义教育基地（一号工程）
延安基地建设项目开工典礼

开工典礼 Project Operation Ceremony

开馆典礼 Openning Operation Ceremony

后记 | Postscript

■ 2008年7月《物华天宝之馆》出版至今转眼已过三年。《长安意匠》这套作品集的第五集迟迟未能问世,其间的情况令我亦喜、亦忧、亦犹豫。喜的是我所属中国建筑工程总公司决定将此作品集纳入公司的科研项目《张锦秋建筑创作道路与思想成果集成研究》课题,因而成果的出版从此有了强大的经济后盾;忧的是出版过程中的种种纠结,为我所不了解,而又明显影响了出书的进程,现终于理清了;犹豫的是五年前拟定《长安意匠》的七个分册名称,似已不能涵盖新形势下近期建成的新项目。分册目录不得不做必要的调整。

■ 现在一切就绪,我们终于在中国共产党建党90周年的前夕,将第五分册《延安革命纪念馆》脱稿付印,以此作为我们的献礼。谢谢编辑、印刷、出版各单位,无论如何让这个分册在七月内问世。

■ It's been three years since the publication of "Museum of Treasures" in July 2008. The fifth volume of "Artistic Conception Practiced in Chang'an" has failed to be published. Meantime, I was happy, anxious and hesitating. The reason for happiness was that the volume was included in the topic of "the Study on Collection of Achievements of Zhang Jinqiu's Architectural Creation Road and Thoughts" by China General Building Construction Company, in which I have been working. Hence publications of the achievements had obtained strong financial support. The reason for anxiousness was that I wasn't familiar with the problems occurred in the process of publication, which obviously had effects on the publishing work. All these problems were finally solved. The reason for hesitation was that the titles of the seven volumes of "Artistic Conception Practiced in Chang'an" could not include the recently completed projects. Alterations on the titles were necessary.

■ Now everything is ready. On the eve of the 90th anniversary of the founding of the Chinese Communist Party, the fifth volume of "Yan'an Revolutionary Memorial Hall" has been completed and put into printing. This is taken as a gift for the celebration of the Party. Herewith I sincerely express my thanks for those in editing, printing and publishing for successful publication of the volume in July of the year.

设计组合影 Design Team

长安意匠——张锦秋建筑作品集

张锦秋 著

承　　编：《建筑创作》杂志社　中国建筑西北设计研究院
主　　编：金　磊　王舒展
执行主编：康　洁
编　　辑：艾学农　吴阳贵　王　军
摄　　影：杨超英　吴阳贵★
英文翻译：刘　河
英文校对：苗　淼　王舒展
书籍设计：康　洁
编　　务：刘　佳　王晓蓉
资料提供：中国建筑西北设计研究院华夏所

★ 图号后加△为吴阳贵摄，加口为延安革命纪念馆提供

- Author: Zhang Jinqiu
- Editorial Execution: *Architectural Creation* Magazine
  China Northwest Building Design Research Institute
- Chief Editor: Jin Lei, Wang Shuzhan
- Executive Chief Editor: Kang Jie
- Editors: Ai Xuenong, Wu Yanggui, Wang Jun
- Photographers: Yang Chaoying, Wu Yanggui
- Translator: Liu He
- Copyreader: Miao Miao, Wang Shuzhan
- Book Designer: Kang Jie
- Editing Assistants: Liu Jia, Wang Xiaorong
- Information Provider: Chinese Architectural Department
  of China Northwest Building Design Research Institute

图书在版编目(CIP)数据

延安革命纪念馆/张锦秋著.—北京:中国建筑工业出版社,2011.7
(长安意匠——张锦秋建筑作品集)
ISBN 978-7-112-13418-2

I.①延… II.①张… III.①纪念馆-建筑设计 IV.①TU251.3

中国版本图书馆CIP数据核字(2011)第140981号

责任编辑:费海玲 张振光
责任校对:刘 钰

长安意匠——张锦秋建筑作品集
My Artistic Conception Practised in Chang'an–Selection of Zhang Jinqiu's Architectural Creation
延安革命纪念馆
YAN'AN REVOLUTION MONUMENT
张锦秋 著

中国建筑工业出版社出版、发行(北京西郊百万庄)
各地新华书店、建筑书店经销
北京雅昌彩色印刷有限公司印刷

开本:787×1092毫米 1/12 印张:10 1/3 字数:320千字
2011年7月第一版 2011年7月第一次印刷
定价:128.00元
ISBN 978-7-112-13418-2
(21168)

版权所有 翻印必究
如有印装质量问题,可寄本社退换
(邮政编码100037)